Great Rescues

Henry Billings

Melissa Stone

STECK-VAUGHN
C O M P A N Y
A subsidiary of National Education Corporation

Books in this series:
Great Disasters
Great Escapes
Great Mysteries
Great Rescues

Acknowledgments

Supervising Editor
Kathleen Fitzgibbon
Project Editor
Christine Boyd
Designer
Sharon Golden
Photo Editor
Margie Matejcik

Illustration Credits
Cover Illustration: Floyd Cooper

Peg Dougherty pp. 48-50, 52-53
Jimmy Longacre pp. 16-18, 72-74, 76

Photo Credits
Pp. 3, 4, 5, Courtesy of U.S. Navy; pp. 8, 9 UPI/Bettmann Newsphotos;
p. 10 Lomen Family Collection/Archives, University of Alaska Fairbanks;
p. 11 © Robert L. Miller/Animals Animals; p. 12 Brown Brothers; pp. 22, 23
Carl Mydans/LIFE MAGAZINE © Time Inc.; p. 24 Joe Scherschel/LIFE
MAGAZINE © 1958 Time Inc.; p. 28 © Chip Porter/Alaska Photo/ALLSTOCK;
pp. 31, 33 UPI/Bettmann Newsphotos; pp. 36, 37, 38, 39 Courtesy of
U.S. Coastguard; pp. 42, 43 UPI/Bettmann Newsphotos; p. 44 © Sygma;
p. 45 © James K.W. Atherton/The Washington Post; pp. 56-57, 58 © Simon
Ward/ALLSPORT; p. 60 © Focus on Sports; p. 61 AP/Wide World; p. 64
© Scott Shaw/Sygma; p. 65 © Judy Malgren/Sygma; p. 67 © M.Rogers/Sipa
Press; p. 68 © David Woo/Sygma; p. 69 © M.Rogers/Sipa Press; p. 81 ©
Schultz/Sipa Press; pp. 82, 83 © Rich Frishman/Picture Group.

ISBN 0-8114-4176-8

4 5 6 7 8 9 0 PO 94 93

Contents

A Ship in Flames 2

Dogsleds to the Rescue 8

Trouble on the Mississippi 16

Miracle at Springhill 22

Survival in the Yukon 28

Abandon Ship! 36

Pulled From the Potomac 42

Do or Die! 48

Free Fall Rescue 56

Trapped in a Well 64

The Little Boy Who Could 72

A Path to Freedom 80

Glossary 86

Progress Chart 92

Answer Key 93

A Ship in Flames

The French ship, Vinh-Long, was sailing off the coast of Turkey. An officer on board smoked a **cigarette** as he kept watch. It was almost dawn, and he was tired. As he left the deck, he dropped his cigarette. It landed near some trash.

Within moments, a fire had started. The flames crept toward a leaking gasoline drum. Suddenly, fire shot into the air. A great **explosion** followed. The Vinh-Long carried a large amount of gunpowder. If the fire set it off, there would be no hope. The huge French ship would explode. All 500 people aboard would be killed.

Help Is on the Way

The fire broke out on December 16, 1922. Three miles away from the Vinh-Long sailed a tiny American ship – the Bainbridge. An officer on the Bainbridge saw the flash of fire. He sent for Captain W. A. Edwards.

When Edwards saw the flames, he sounded an alarm. Sailors dashed to their stations. They grabbed fire and rescue **equipment**. Then they lined up by the lifeboats. They were ready for action. Edwards smiled. He knew he had a good crew.

Soon the Bainbridge reached the Vinh-Long. Edwards yelled out, "Ship ahoy! Can we be of **assistance**?"

The French captain was about to answer when another blast shook the Vinh-Long.

Edwards later said, "His ship answered for him. A terrific roar deafened us. The middle of his ship had blown up. I knew a **dreadful experience** lay ahead."

The force of the blast had blown people into the water. Screams of frightened people filled the night. The Bainbridge sailors rowed out into the dark waters. They lifted people out of the water as fast as they could.

Danger Ahead

One of the people rescued was a French sailor. He was half out of his mind with fear. "She's loaded with gunpowder," he screamed. "She will blow up in a moment! You will be blown up, too!"

Captain Edwards stopped to think. His ship also carried **explosives**. One well-placed spark and his ship would be blazing just like the Vinh-Long.

Edwards said later, "I wanted to help the people on the blazing ship. But I could not forget that I had

Sailors rescued from the Vinh-Long

nearly one hundred fine young Americans on my own ship. They were ready to obey my every order. The sight of dying men did not make them **flinch**. But I flinched when I thought of asking them to do what seemed the only right and decent thing to do."

The captain didn't want to put his men in danger. But he had to do something. He brought his ship next to the Vinh-Long. The people on the Vinh-Long crowded against the rail. They got ready to jump onto the Bainbridge. They knew the huge ship could blow up at any moment.

Just then came the worst explosion yet. People on both ships were knocked off their feet. Everyone was blinded for a moment. The force of the blast pushed the Bainbridge away from the Vinh-Long.

President Coolidge presents Medal of Honor to Captain Edwards.

Stand by to Ram!

Edwards knew he had to try something else. He would **ram** the French ship. He hoped this would damage the Vinh-Long's **hull**. Then water would rush in and slow down the fire. If his plan worked, the people on board would have time to get off. If it didn't work, both ships might explode.

Edwards gave the order. The Bainbridge rammed into the Vinh-Long with perfect aim. The flaming ship's hull broke open. The people on board went wild. They knew that every second counted. They poured onto the Bainbridge. Soon everyone who was still alive was on the American ship.

Edwards ordered his ship to move away. The Bainbridge pulled free. The rescue was complete! Edwards and his crew had saved 482 people. President Coolidge later presented Captain Edwards with the Medal of Honor for his brave acts.

Do You Remember?

■ In the blank, write the letter of the best ending for each sentence.

_____ 1. The Bainbridge was a
   a. big French ship.   b. small U.S. ship.   c. lifeboat.

_____ 2. The fire was started by
   a. lightning.   b. W. A. Edwards.   c. a cigarette.

_____ 3. The Vinh-Long carried
   a. gunpowder.   b. no passengers.   c. circus animals.

_____ 4. Some people got off the Vinh-Long by jumping onto
   a. helicopters.   b. the Bainbridge.   c. gasoline tanks.

_____ 5. The Bainbridge finally
   a. sank.   b. blew up.   c. crashed into the Vinh-Long.

Express Yourself

■ Imagine you are a crew member of the Bainbridge. Write a letter to a friend about your experience with the rescue of the Vinh-Long.

Dear _____,

_____

_____

_____

_____

_____

_____

_____

Exploring Words

■ Choose the correct word from the box to complete each sentence.

| cigarette | explosion | equipment | assistance | dreadful |
| experience | explosives | flinch | ram | hull |

1. The sailors grabbed fire and rescue _____.

2. When you help someone, you give _____.

3. A _____ is a roll of tobacco that is wrapped in paper and smoked.

4. When the gunpowder caught on fire, there was a great _____.

5. The body of a ship is called the _____.

6. Something that is _____ is very bad.

7. Materials used to blow something up are called _____.

8. If you _____, it means that you back away from danger.

9. An _____ is something you see, do, or live through.

10. To run into something with great force is to _____ it.

Dogsleds to the Rescue

Dr. Curtis Welch stared at the sick young boy. He had a **fever**, and his throat was **swollen**. Every few minutes he broke into a deep cough.

Dr. Welch shook his head and muttered unhappily, **"Diphtheria."**

"What does that mean? Is he going to die?" asked his helper.

"He and many others will if we don't get medicine soon. Diphtheria spreads very quickly," Dr. Welch replied.

A Call for Help

Dr. Welch was worried. Nine out of ten people who came down with diphtheria died. There was a cure—a medicine called **antitoxin**. But Welch didn't have any. There wasn't any for miles around. He feared the **disease** would kill everyone in the frozen town of Nome, Alaska.

Dr. Welch met with town **officials**. They made a plan to check the spread of the disease. Schools were closed. The movie house was also closed. No one knew how many people might have the disease. The less people were around each other, the less chance they would have of catching diphtheria.

The next day, January 27, 1925, Dr. Welch sent a message to other Alaskan cities. He needed antitoxin. And he needed it fast.

The nearest antitoxin was found in Anchorage, about 900 miles away. But would it arrive in time? All the roads were closed for the winter. Planes couldn't fly because of the high winds and cold temperature. The railroad tracks only came as far as Nenana, which was over 650 miles from Nome.

Dr. Welch didn't know how the medicine could be brought to Nome. But he did know that without it, many of the 1,429 people who lived in Nome would die.

The governor of Alaska had to find a way to get the antitoxin to the small town. He decided that it should be sent by railroad from Anchorage to Nenana. From there, dog teams would carry it into Nome. The plan had to be carried out quickly, or it would be too late.

The Brave Ones

Leonard Seppala

In Anchorage, workers packed the antitoxin. They sent it by train to Nenana. From there, it would have to be carried by **relay** teams of dogsleds. A **musher** and his dogs would take the antitoxin from Nenana to the next town. The next musher would then take the antitoxin. He and his dogs would carry the medicine for the next leg of the trip. The antitoxin would change hands about 20 times before it reached Nome.

The trip would be difficult. The mushers would lead their dogs across the icy wilderness. They would face –50° F temperatures, howling winds, and blinding snow. Some of the dogs might not make it. Some of the mushers might not either.

Many mushers quickly stepped forward to do the job. They knew it was dangerous. But these were

brave and rugged people. They had spent their lives in the frozen lands of the North. They knew how to handle their dogs. They knew the route between Nenana and Nome. The mushers were willing to **risk** their lives to help the people of Nome.

One by one, the mushers carried the antitoxin west toward Nome. Upon meeting, the mushers would exchange the antitoxin, instructions for its care, and news about the spread of the disease. All of Alaska watched and waited during this race against time and the weather.

A Dangerous Short Cut

On February 1, the antitoxin reached the small village of Shaktolik. Shaktolik was still over 150 miles from Nome. Here musher Leonard Seppala made a bold decision.

"I'm not going to follow the trail up around Norton Bay," he said. "I'm going to take the short cut straight across the frozen water."

"In this storm?" one of the townspeople asked. "Those 80-mile-per-hour winds will break up the ice. You and your dogs will fall in and be swallowed up by the water."

But Seppala knew that if he made it, he would save precious hours. All around him the snow blew in angry clouds. His dogs slipped on the ice. They groaned. They whimpered. But still he drove them forward. At last, he reached the other side of the bay. He passed the antitoxin on to the next musher.

Gunnar Kasson and Balto

A Near Miss

Gunnar Kasson was the last musher. He had to travel 60 miles to reach Nome. It was 8:00 P.M., and the sky was black. But he couldn't wait for morning. The people in Nome were waiting. Dr. Welch had already reported that five people had died from diphtheria. And at least 30 more people had caught the disease.

Kasson called to Balto, his lead dog. Balto guided the sled to the snowy trail. Soon, however, the wind grew worse. In his 22 years of driving dogsleds, Kasson had never seen it this wild. The freezing air cut through his heavy winter clothes. It froze his right cheek. His hands ached. The dogs too were in pain. Pieces of ice stuck in their feet. Their paws began to bleed.

And then the worst thing of all happened. Kasson lost the trail. The swirling snow had swept away his sense of direction. He called out to Balto. He prayed that Balto could pick up the scent of the trail again. If not, all was lost. Balto sniffed around in the snow. He turned one way, then another. After several anxious minutes, Balto picked up speed. He had found the trail!

Finally, at 5:36 the next morning, Kasson and his dogs limped into Nome. Kasson brought the sled to a stop. Then he **collapsed** in the snow next to his half-frozen dogs.

"Fine dog," he mumbled again and again. "Balto, you brought us through. You brought us through."

People cheered all the dogs who had made the trip. Kasson and Seppala and the other mushers were heroes. These brave people and their animals had saved the lives of many Alaskans.

Do You Remember?

■ Read each sentence below. Write **T** if the sentence is true. Write **F** if the sentence is false.

_____ 1. Dr. Welch was worried.

_____ 2. The nearest antitoxin was in Montana.

_____ 3. Nome was a very large city.

_____ 4. One man would carry the antitoxin the whole way.

_____ 5. The journey was dangerous.

_____ 6. Seppala took a short cut to save time.

_____ 7. Kasson was the first musher.

_____ 8. Balto found the trail in the snowstorm.

Critical Thinking — Cause and Effect

■ Complete the following sentences.

1. When Dr. Welch saw that the child had diphtheria, he was very

   worried because _____

   _____

2. The antitoxin was sent to Nome from Anchorage because _____

   _____

3. Planes couldn't fly because _____

   _____

4. The dogs' paws began to bleed because_____

   _____

Exploring Words

■ Read each sentence. Fill in the circle next to the best meaning for the word in dark print. If you need help, use the Glossary.

1. The word **fever** means
   ○ a. high body temperature.   ○ b. pain.   ○ c. not as many.

2. The word **swollen** means
   ○ a. a river.   ○ b. larger than normal size.   ○ c. sore.

3. The word **diphtheria** means
   ○ a. a sickness.   ○ b. a game.   ○ c. a religion.

4. The word **relay** means
   ○ a. a kind of race.   ○ b. a smile.   ○ c. slippery.

5. The word **officials** means
   ○ a. losers.   ○ b. rich people.   ○ c. people in charge.

6. The word **musher** means
   ○ a. a meal.
   ○ b. a police officer.
   ○ c. a person who drives a dogsled.

7. The word **disease** means
   ○ a. gasoline.   ○ b. a small animal.   ○ c. an illness.

8. The word **risk** means
   ○ a. tear apart.   ○ b. take a chance.   ○ c. remove.

9. The word **antitoxin** means
   ○ a. yelled.   ○ b. a medicine.   ○ c. dusty.

10. The word **collapsed** means
    ○ a. fell down.   ○ b. waited.   ○ c. relaxed.

Trouble on the Mississippi

"Tom, bring the <u>Zev</u> up the river from Helena. Don't pile it up on a **sandbar**. And keep an eye on floating logs," Mr. Hunter ordered.

Tom Lee nodded his head at his boss. He would take the motorboat, <u>Zev</u>, up the Mississippi River. He would take it from Helena, Arkansas, to Memphis, Tennessee.

But this would not be an ordinary trip. By the end of the day, Tom Lee would become a hero. He would save 32 people from **death**.

A Happy Cruise

On May 8, 1925, the river looked peaceful. But Lee knew it could be tricky. The spring flood **currents** were dangerous. But he wasn't worried. He knew how to handle the boat. And he could deal with the currents.

As the Zev moved along, it passed a large steamboat. The steamboat was called the M.E. Norman. Lee saw that it was out for a **cruise**. The men and women on board wore their best clothes. They seemed to be having a good time. Several people waved to Lee. He waved back.

Lee passed the Norman. Then he glanced back. He noticed something strange about the steamboat.

"That boat is riding funny," he said to himself. "She's rolling too much to one side to suit me. I'll keep an eye on her."

Trouble!

Lee had pulled about a half a mile ahead of the steamboat. Then it happened. The Norman started to roll crazily. The sun had just gone down. A **mist** had settled over the river. Lee couldn't see the Norman very well. But he knew it was in trouble. Quickly, he turned the Zev around and raced back down the river.

The steamboat was already on its side and sinking by the time he reached it. Lee could see heads **bobbing** in the water. He pushed the Zev's motor to full power.

It was getting darker by the minute. And the Norman was far from shore. Lee had no time to waste.

As Lee got closer to the steamboat, it sank out of sight. Lee could see people splashing in the water. He heard cries for help. The swift current was carrying people down the river. Luckily, the <u>Zev</u> had a **powerful** motor. Lee was able to act in a hurry.

Saving Lives

First, he had to get down river from the **struggling** swimmers. If he didn't, they would float out of reach. Lee steered the <u>Zev</u> past them. Then he turned around and came back up the river.

Tom Lee moved the <u>Zev</u> slowly through the dark water. Again and again he stopped and pulled people into his boat. Soon the <u>Zev</u> was full. Lee brought the frightened people to the shore 300 feet away. Some were in **shock** and had to be lifted from the boat onto the ground.

Then Lee went back to find others. Luckily, some of the passengers were strong swimmers. Seventeen of them managed to swim to the shore safely. Others held on to **life preservers**. Still others held on to pieces of wood. All were drifting down the river.

Lee got as many as he could. He made five trips between the shore and the river. He saved 32 people. Lee helped them make a fire on shore. Then he spent the rest of the night searching the river. He was trying to find more people. In all, 23 drowned.

James Wood was one of the 32 passengers saved by Tom Lee. "If it had not been for Tom Lee, I would not be here today. I owe my life and my wife's life to him. And there are many others who will say the same thing."

When asked about the rescue, Lee said he just happened to pass the sinking boat. He said he did what anyone else would have done in his place. But to the people on the <u>Norman</u>, Tom Lee was not just an ordinary person. He was a brave and skillful man who had meant the difference between life and death.

Do You Remember?

■ Read each sentence below. Write **T** if the sentence is true. Write **F** if the sentence is false.

_____ 1. Lee's boss told him to take the <u>Zev</u> up the river.

_____ 2. Lee had never been out on the Mississippi before.

_____ 3. The <u>Zev</u> passed by the <u>Norman</u>.

_____ 4. Some people on the <u>Norman</u> were strong swimmers.

_____ 5. The <u>Zev</u> was a steamboat.

_____ 6. Lee quit looking for people when it got dark.

_____ 7. Thirty-two people were saved by Lee.

_____ 8. Tom Lee drowned while rescuing James Wood.

Express Yourself

■ Imagine that you are a reporter writing about the boat accident and Tom Lee. Tell who, what, where, when, and why in your newspaper article.

_____

_____

_____

_____

_____

_____

_____

Exploring Words

■ Use the clues to complete the puzzle. Choose from the words in the box.

bobbing
cruise
currents
death
life preservers
mist
powerful
sandbar
shock
struggling

Across

2. when a body shuts down after an injury or accident
5. very strong
8. moving bodies of water
9. a light rain
10. moving up and down

Down

1. things used to keep people from drowning
3. a boat ride taken for fun
4. trying hard
6. the end of life
7. ridge of sand on the bottom of a river

Miracle at Springhill

The **miners** of Springhill knew the risks. Mining was tough, dangerous work. Over the years, hundreds of men had died in the mines. Others became sick from breathing coal dust. But the small Canadian town of Springhill was a mining town. If a young man wanted a job, he worked in the mines as his father and grandfather had.

On October 23, 1958, the earth shook **violently**. This sudden movement of the earth was called a "bump."

Giving Up Hope

This bump trapped 174 men in the mines. Eighty-one were quickly rescued. Twenty-four others were known dead. That meant the remaining 69 men were still trapped 4,000 feet straight down.

For the next six days, rescue workers tried to dig down to the trapped miners. But it was a very deep mine. And the digging went slowly.

Family and friends waited day after day. They stood around the entrance to the mine. They prayed for good news. But none came. Slowly, they began to lose hope. By the morning of October 29, most people had stopped coming to the mine. Families started planning church services for their lost men.

Rescue workers kept digging. But by the sixth day, they had no real hope of finding anyone alive.

Rescue workers hurry to the mine.

On the afternoon of October 29, a rescue worker uncovered the top of a 60-foot pipe. He heard a voice. It was coming from the far end of the pipe!

"There are twelve of us in here. Come and get us."

Staying Alive

The twelve men were stuck in a space only four feet high. They had not eaten food or breathed fresh air for days. Yet somehow, they were still alive.

Three of the miners had lived through a bump two years before in 1956. They had been trapped for three days. These men knew what to do. They helped the others stay **calm**.

The twelve men searched for food. They found a few lunch boxes that contained water and sandwiches. They tried hard to keep their **spirits** up. They told jokes, sang songs, and prayed aloud.

After three days, the food was gone. So was most of the water. The air grew **stale**. Eldred Lowther said, "I thought I was lucky by not being killed in the bump. But by the fourth day, I began to think it would have been better if I had died. I lost all hope."

Every few hours the miners thought they heard rescue workers. They yelled for help. But they never got an answer. They found an air pipe and took turns tapping out **SOS** on it. But they feared the pipe was broken and their efforts useless.

Miracle at Springhill

After six days, the miners were weak. They were terribly hungry and thirsty. But they kept tapping on the pipe. Suddenly, they heard a sound from the other end of the air pipe. Tears formed in the miners' eyes. They were going to be saved!

Up above, the good news flashed through the town. Doctors and more diggers rushed to the mine. Family and friends also came running.

The rescue workers dug quickly. They got within 20 feet of the miners. But they found they could go no farther.

"We have to go back out and start again," one of the diggers shouted down the pipe.

But it was only a short **delay**. The digging went on. At last, after thirteen hours, rescue workers reached the **survivors**. In a short time, all twelve men were safely out of the mine.

Two days later the rescue workers found seven more miners alive. All the **buried** miners had been given up for dead. But nineteen of them had survived. To the people of Springhill, this rescue really was a **miracle**.

A survivor is brought up from the mine.

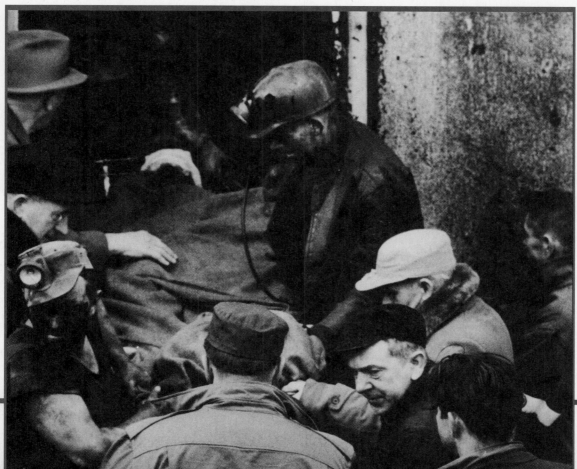

Do You Remember?

■ In the blank, write the letter of the best ending for each sentence.

_____ 1. For the first few days, families of the trapped men waited
a. near the mine.　　b. at church.　　c. in a hotel.

_____ 2. The miners tapped out SOS on
a. a rock.　　b. an air pipe.　　c. a lunch box.

_____ 3. Three of the trapped miners had lived through an earlier
a. bump.　　b. hurricane.　　c. flood.

_____ 4. While trapped in the mine, the miners told
a. ghost stories.　　b. jokes.　　c. tall tales.

_____ 5. After six days, the miners were
a. wet.　　b. dead.　　c. rescued.

Critical Thinking — Fact or Opinion?

■ A **fact** can be proven. An **opinion** is a belief. Opinions cannot be proven.

Write **F** before each statement that is a fact. Write **O** before each statement that is an opinion.

_____ 1. If a young man wanted a job in Springhill, he worked in the mines.

_____ 2. Being a miner is no fun.

_____ 3. Some miners were killed in the bump on October 23, 1958.

_____ 4. Friends of the trapped miners gave up hope too soon.

_____ 5. This mine was very deep.

_____ 6. The trapped miners sang songs.

_____ 7. The twelve miners should have tapped on the pipe more often.

_____ 8. The twelve men all made it out of the mine alive.

Exploring Words

■ Choose the correct word from the box to complete each sentence.

violently	stale	spirits	buried	SOS
survivors	miracle	miners	delay	calm

1. When something is no longer fresh, it is _____.

2. A worldwide signal for help is _____.

3. Something that is under the ground is _____.

4. People who work in mines are called _____.

5. When people are sad, their _____ are low.

6. People who live through some kind of test or trial are called

   _____.

7. When something happens with great force, it happens

   _____.

8. Something that is quiet and peaceful is _____.

9. A wonderful event that can't be explained is called a

   _____.

10. To put an event off to a later time is to _____ it.

Survival in the Yukon

Helen Klaben stared out the window of the small plane on February 4, 1963. The snowstorm was getting worse. It was hard to see where they were going. They were flying from Fairbanks, Alaska, to Seattle, Washington. But pilot Ralph Flores wasn't sure where they were anymore. He dropped to a lower **altitude**, hoping to get a clue from the ground.

But this part of Northern Canada, called the Yukon, had no towns. It had only huge, deep forests. Ralph put down his hand to change gas tanks. When he looked up, the plane was headed right for the trees.

Things Look Bad

In the next instant, the plane crashed. It fell through the trees and landed in the snow. All was silent. Ralph lay **slumped** against the instruments. Helen, the only passenger in the plane, lay in her seat without moving.

Some time later, Helen woke up. She looked over at the pilot.

"Are you all right, Ralph?" she asked.

Ralph moved a bit. "I'm all right," he replied.

But really, neither Ralph nor Helen was all right. Blood poured from Ralph's mouth. His jaw was broken. His ribs were **smashed**. His hands and feet suffered from **frostbite**. Helen suffered from frostbite, too. Her chin was cut open. And her left arm was badly broken.

It seemed certain they would die. There was no one to help them. The nearest road was 60 miles away. The wind howled, and the snow drifted down through the trees. The temperature was −48° F.

Helen and Ralph had lost their boots in the crash. They had no sleeping bags, blankets, or **waterproof** clothing. They had very little food – some crackers, a little fruit, and six small cans of fish.

But Ralph and Helen did not give up. They didn't want to die. And so they began the long, hard fight to stay alive in this **hostile** place.

Staying Alive

Helen found some matches in her pocket. Ralph collected wood. They made a small fire next to the plane. They found extra clothes in Helen's suitcase. They put on as many layers as they could. Then they tied sweaters around their feet.

This message was left on the plane by Ralph.

Slowly, they cleared out the back of the broken plane. They made a shelter from the cold and wind. That night, they curled up next to each other and tried to sleep. The next day they were hungry. But they ate very little. They didn't know how many days they would be here. And they wanted to make the food last as long as possible.

Every day they stared at the sky, watching for rescue planes. Both of them were in pain. Helen's broken arm ached. Her toes turned black with frostbite. Ralph's jaw and chest **throbbed**. Hunger knotted their stomachs.

By the tenth day, they ran out of food. Now all they had to eat was melted snow and two tubes of toothpaste. Helen's frostbitten toes made it difficult for her to walk. She spent most of her time in the shelter. She read to keep her spirits up.

Ralph stayed busy working. He cut wood for the fire. He worked on the plane's broken radio. He made a flag and tied it to the top of a tree. He even tried to build a trap to catch rabbits.

Many times they heard planes flying overhead. A few times they even saw planes. They built fires and flashed mirrors. They tried to make radio **contact**. But none of the planes noticed them.

Searching for Help

By the 32nd day, they gave up hope that anyone would find them in the forest. They had to get to a clearing where they could signal passing planes. Ralph set out to scout the area. He told Helen he would be back in three days.

About a mile away, he found a clearing. So he built a shelter and waited. He waited for eight long days, but no planes came.

Finally, he couldn't wait any longer. He was worried about Helen. He wanted to make sure she was all right, so he returned to the plane.

Helen was so happy to see Ralph that tears ran down her cheeks. They had now been in this frozen forest for 40 days.

They knew they had to try again to get help. This time Helen would also go. Her feet were still in bad shape from frostbite. So Ralph spent a few days making a sled for her to ride on. Then he began dragging it down a hill.

The sled kept tipping over, so finally Helen got out and walked. The snow came up to her waist. She lost the sweaters that had been wrapped around her feet. She was sick to her stomach. Yet, she found the strength to keep going.

Ralph's whole body ached. He had stomach **cramps**. But somehow he kept moving.

At last they came to a large clearing. Ralph made a fire and boiled snow to make water for them to drink. He built a new shelter out of tree branches. Then he and Helen lay down to sleep. They were **exhausted**.

Several days later, Ralph decided that the clearing was not big enough. He wanted to find a larger field. But Helen didn't think she could go any farther. So Ralph took off by himself. Slowly, he traveled four miles to a huge meadow.

There he stamped out the letters SOS in the snow. He made the letters as large as he could. Then he made an arrow that pointed toward Helen's shelter. A bush pilot named Chuck Hamilton saw the letters. He called the police. And so, 49 days after the crash, rescuers found Ralph and Helen.

Some rescue workers were shocked that Helen and Ralph were still alive. They didn't think that anyone could last more than two weeks in such hostile conditions. But both Helen and Ralph had the will to live. They helped each other, and they never gave up.

Helen Klaben in the hospital after the rescue

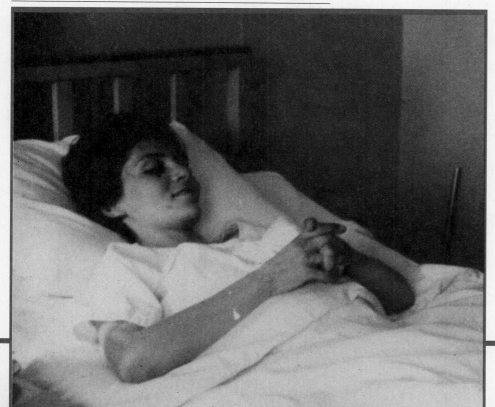

Do You Remember?

■ In the blank, write the letter of the best ending for each sentence.

_____ 1. Helen and Ralph's plane crashed in a
a. lake.   b. large meadow.   c. forest.

_____ 2. In the crash, Helen hurt her
a. back.   b. left arm.   c. right eye.

_____ 3. After ten days, Helen and Ralph ran out of
a. firewood.   b. food.   c. money.

_____ 4. Helen and Ralph often heard
a. gunshots.   b. planes.   c. laughter.

_____ 5. A pilot finally spotted Ralph's
a. SOS.   b. smoke signals.   c. jacket.

Express Yourself

■ Pretend that you are Helen Klaben. Your plane has crashed in the Yukon. It has been 40 days, and still you have not been rescued. Write an entry in your journal describing your thoughts, hopes, and fears.

_____

_____

_____

_____

_____

_____

_____

Exploring Words

■ Use the words in the box to complete the paragraphs. Reread the paragraphs to be sure they make sense.

altitude	**smashed**	**exhausted**	**slumped**	**throbbed**
frostbite	**hostile**	**waterproof**	**cramps**	**contact**

The small plane flew blindly through the clouds. Ralph dropped to a lower **(1)** _____ to see better. Then the plane **(2)** _____ into some trees. When the plane stopped moving, Ralph and Helen were **(3)** _____ in their seats. They were stuck in a cold, **(4)** _____ wilderness.

As the days passed, they grew weaker. They were cold and wet. They had no **(5)** _____ clothing. Helen's toes turned black with **(6)** _____. Ralph's jaw **(7)** _____ with pain. They both had stomach **(8)** _____ from hunger. They became **(9)** _____ as they made their way to an open field. Finally, Ralph made **(10)** _____ with a bush pilot by stamping SOS in the snow. After seven long weeks, they were finally rescued.

Abandon Ship!

"This is your captain speaking. There is a small fire in the engine room. It is under control. There is no cause for alarm. But for your own safety, please report to the upper deck."

That message came about 1 A.M. on the morning of October 4, 1980. Most of the 324 passengers were not worried. After all, the Prinsendam was a safe ship. So why should they **panic?** Some passengers took their life preservers. Others didn't bother. They went to the upper deck dressed in their pajamas.

Fire!

The Prinsendam left Vancouver, Canada, on September 30. It was the beginning of a month-long luxury cruise to Asia. Most of the passengers were 65 years old and older. Many had saved for years to take this trip. But after just three days at sea, **disaster** struck.

The trouble began 120 miles off the coast of Alaska. The **fuel system** started to leak. Oil from the engine **spurted** onto a hot pipe. It burst into flames. Crew members sealed off the engine room. They tried to **smother** the flames. But the fire kept **raging**.

At 2:25 A.M., the Prinsendam sent out an SOS. **Admiral** Schoel at the Coast Guard Center in Juneau, Alaska, received the message. He radioed officials in Kodiak Island, Anchorage, and British Columbia. Soon, **helicopters** were flying toward the Prinsendam from all directions.

The ship's SOS had also been heard by the Williamsburgh. The Williamsburgh was a huge oil tanker. It was 90 miles south of the Prinsendam. It turned around and sailed toward the troubled ship.

Meanwhile, the Prinsendam crew kept fighting the fire. They did the best they could. But at 4:54 A.M., Captain Cornelius Wabeke saw it was no use. He announced the news over the loudspeaker.

"I'm sorry. The fire is completely out of control. We have to **abandon** ship."

Into the Boats

Getting the lifeboats loaded and into the water was not easy. The fire had cut off the ship's electricity. There were no lights. Finally, a helicopter arrived. It pointed a bright light on the ship.

A lifeboat full of passengers

By 6:30 A.M., all the lifeboats and their passengers were lowered into the water. Captain Wabeke worried as he watched the lifeboats leave. He wondered how the older passengers would manage. Some of them had weak hearts. A few were in wheelchairs. He hoped the sea would remain calm.

Luckily, the Williamsburgh, the huge oil tanker, had arrived. One of the lifeboats got next to the tanker. Then someone dropped a rope ladder over the side.

One by one, the frightened passengers tried to climb the rope ladder. But it was a long 40-foot climb. It would take more than a full day to rescue all 524 people this way. And a storm was blowing in. The waves were getting higher. The other lifeboats were drifting away from the Williamsburgh. Something else needed to be done.

A helicopter lifts people from the sinking Prinsendam.

The Biggest Sea Rescue Ever

More helicopters arrived. An order was given to lift people from the lifeboats onto the Williamsburgh. The pilots lowered one-person baskets to the lifeboats. The pilots had to keep their helicopters steady. The baskets swung in the wind.

As a basket was lowered, someone grabbed it and crawled in. The passenger held on tightly as the basket was raised to the helicopter. Some people were so scared they kept their eyes shut.

When a helicopter was full, it would fly to the Williamsburgh. There passengers would be unloaded. The helicopter would then return to pick up more people from the lifeboat. Crew members made the trip again and again.

Finally, by the next morning, the rescue was complete. All 524 people on board the Prinsendam had been saved. None had been badly hurt. The largest sea rescue in history was a success.

Do You Remember?

■ Read each sentence below. Write **T** if the sentence is true. Write **F** if the sentence is false.

_____ 1. The Prinsendam was sailing to Asia.

_____ 2. Captain Wabeke gave the order to abandon ship.

_____ 3. Only a few passengers made it into lifeboats.

_____ 4. The Prinsendam sent out an SOS.

_____ 5. The crew of the Williamsburgh refused to help the passengers of the Prinsendam.

_____ 6. Helicopters joined in the rescue effort.

_____ 7. Passengers were rescued by motorboats.

_____ 8. All 524 people on board were saved.

Critical Thinking — Drawing Conclusions

■ Finish each sentence by writing the best answer.

1. The fire broke out because _____

_____

2. The passengers were told to abandon ship because _____

_____

3. The Williamsburgh turned around because _____

_____

4. Captain Wabeke was worried about the passengers because _____

_____

_____

Exploring Words

■ Read each sentence. Fill in the circle next to the best meaning for the word in dark print. If you need help, use the Glossary.

1. People did not **panic** when they first heard about the fire.
   ○ a. become frightened   ○ b. cheer   ○ c. listen

2. Oil from the engine **spurted** onto a hot pipe.
   ○ a. shined   ○ b. poured out   ○ c. planted

3. The cruise turned into a **disaster.**
   ○ a. mystery   ○ b. terrible failure   ○ c. great joke

4. **Fuel** from the engine caught on fire.
   ○ a. oil and gas   ○ b. sound waves   ○ c. food

5. The lighting **system** did not cause the fire.
   ○ a. new worker   ○ b. group of parts   ○ c. disease

6. Crew members tried to **smother** the fire.
   ○ a. find   ○ b. light   ○ c. put out

7. A **raging** storm made the passengers cold and wet.
   ○ a. violent   ○ b. pretty   ○ c. slow-moving

8. **Helicopters** lowered baskets to the lifeboats.
   ○ a. aircrafts   ○ b. volunteers   ○ c. big ships

9. The passengers had to **abandon** the ship.
   ○ a. watch   ○ b. rock   ○ c. leave

10. An **admiral** was in charge of the rescue.
    ○ a. police officer   ○ b. navy officer   ○ c. principal

Pulled From the Potomac

On January 13, 1982, a huge snowstorm hit Washington, D.C. The airport shut down. Planes waited while snowplows cleared the **runways**. Two hours later the airport reopened.

Shortly after that, Air Florida's Flight 90 roared down the runway on its way to Tampa. Joseph Stiley was on that plane. He was a pilot himself, and he **sensed** trouble. The plane didn't climb into the sky the way it should have. Stiley turned to the woman sitting next to him. He said, "We're not going to make it. We're going in."

The Crash

Many workers had been **released** early that afternoon because of the snowstorm. At 4:00 P.M., the streets were **jammed** with people going home. Traffic was at a standstill on the 14th Street Bridge.

Suddenly, a blue and green plane dropped out of the sky. Lloyd Creger was one of the people on the bridge. He watched in **horror** as the plane dropped. "It was falling from the sky, coming right at me," said Creger. "It hit the bridge and just kept on going like a rock into the water."

Leonard Skutnik was a 28-year-old office **clerk**. He was in one of the cars on the bridge. Skutnik and many other people got out of their cars and rushed to the scene.

The plane's tail had ripped the tops off several cars. Four people sitting in cars had been killed. The plane itself had **plunged** into the icy Potomac River.

Skutnik was **stunned**. He and the others watched in silence as the plane disappeared. Only the tail stayed above the water. Of the 79 people aboard, only six could be seen. These six held on to the tail with all their strength. They had to be rescued fast. Some of them were badly **injured**. No one could stay in the freezing water long without dying.

To the Rescue

A rescue helicopter arrived and went to work. It lowered a ring to the water. One person grabbed it and was dragged to the shore. Then a second person was pulled to safety.

Next it was Priscilla Tirado's turn. She was weak and half-frozen. She could hardly reach for the ring. A man in the water next to her helped her. It was Stiley, the **off-duty** pilot. He was one of the six who had survived the crash. He was in trouble, too. Both his legs were broken. Still, he managed to help Tirado catch the ring.

Tirado tried to hold on as the helicopter dragged her out of the water. But she was too weak. The ring slipped from her hands. She fell back into the icy river.

Leonard Skutnik and his family

An Overnight Hero

Back on shore, Leonard Skutnik and the others stood watching. They saw Tirado fall into the water. Skutnik stripped off his shoes and jacket. He plunged into the river. He swam out 25 feet to Tirado. She was so weak and cold she couldn't struggle anymore. Skutnik grabbed her and swam back to shore. When he got close to the bank, a firefighter swam out to help him. Together they carried Tirado to safety.

Skutnik became a hero overnight. His rescue was shown again and again on the news.

But Skutnik didn't see himself as a hero. He said, "Nobody else was doing anything. I just did it."

In the end, five of the six people in the water were saved. Joseph Stiley survived. So did one flight attendant. And because of Leonard Skutnik's brave act, so did Priscilla Tirado.

Do You Remember?

■ In the blank, write the letter of the best ending for each sentence.

_____ 1. Air Florida Flight 90 crashed in
   a. the summer.   b. Alaska.   c. the winter.

_____ 2. The plane fell into
   a. a park.   b. a river.   c. an office building.

_____ 3. When the crash occurred, Leonard Skutnik was in
   a. the plane.   b. a boat.   c. his car.

_____ 4. Priscilla Tirado had trouble hanging onto
   a. her hat.   b. the ring.   c. a police officer.

_____ 5. A ring was lowered into the water by a
   a. firefighter.   b. helicopter.   c. doctor.

Express Yourself

■ Pretend that you are Joseph Stiley. Write an article for a magazine. Describe your experiences on January 13.

_____

_____

_____

_____

_____

_____

_____

_____

Exploring Words

■ Use the words in the box to complete the paragraphs. Reread the paragraphs to be sure they make sense.

runway	plunged	jammed	horror	off-duty
sensed	injured	released	clerk	stunned

Air Florida Flight 90 roared down the **(1)** _____. A few

moments after takeoff, it crashed. It hit a bridge **(2)** _____

with cars and then **(3)** _____ into the Potomac River.

People who saw the plane go down were **(4)** _____.

They watched in **(5)** _____ as the plane sank.

Only six people lived through the crash. One of the six was

Joseph Stiley, an **(6)** _____ pilot. Stiley was badly

(7) _____. Still, he helped Priscilla Tirado grab a

rescue line. But she was too weak to hold the line. She

(8) _____ it and dropped back into the water. Back on

shore, an office **(9)** _____ named Leonard Skutnik

(10) _____ that she was in trouble. Skutnik jumped

into the water and rescued her.

Do or Die!

ire!"

It was early in the morning on June 13, 1982. Karen Hartsock stirred in her bed. She thought she heard her father's voice. A few seconds passed. Then she heard it again.

"Fire! Fire!"

This time Karen was wide-awake. It really was her father's voice. She wasn't having a dream. The house was on fire!

A Young Hero

Fourteen-year-old Karen jumped out of bed. She ran from her second-floor bedroom into the hall. Thick smoke filled the air. Fire was everywhere. Flames covered the floor and the walls. Karen was frightened. Her parents were downstairs where they could escape. But what about her younger brother and two younger sisters? Karen yelled to them. They had to get out of the house fast. Fire was racing through the log farmhouse. Soon they would all be trapped inside.

Eleven-year-old Norma ran crying from her room. She was too frightened to know what to do. Karen grabbed her hand and ran for the stairs. The stairs were already on fire, but this was the only path to safety. Karen pulled her sister through the flames. Then she pushed Norma toward her father. Mr. Hartsock was quite weak. He had just had an **operation** on his heart. But he put out the flames on Norma's pajamas as quickly as he could.

Another Trip

Karen's 12-year-old sister, Loretta, was still upstairs. So was 9-year-old Johnny. Karen would not leave the house without them. She was **especially** worried about Johnny. He had **cerebral palsy** and couldn't get out of the house by himself.

Karen raced back up the blazing stairs. As she did, her nightgown burst into flames.

"Help me!" she cried out in pain.

Fire burned her back and legs. The pain was terrible. But she did not turn back.

She dashed into Johnny's room. At first, she couldn't see anything. The smoke was too thick. Then a flash of fire lit up the room. She saw her little brother still asleep in his bed.

Quickly, Karen wrapped Johnny in a blanket. Then she picked him up and ran back toward the stairs. Burning wallpaper fell on her arms and shoulders. Her hair caught on fire. With one arm, she held Johnny. With the other, she tried to put out the fire in her hair.

A Job to Do

Karen didn't think she could stand the pain. Then suddenly it was gone. She couldn't feel anything. The fire had burned deep into her skin. It had damaged her nerves. She almost felt normal again. At the bottom of the stairs, Karen saw her father.

"Here!" she shouted, throwing Johnny to him.

"I've got him! Now you come on! Come on!" begged her father.

But Karen was thinking about Loretta. She turned and ran back up the burning stairs. Karen didn't know it at the time, but Loretta had already gotten out of the house. When she was halfway up to the second floor, the stairs collapsed. Karen was buried in a pile of burning wood.

Using all his strength, Mr. Hartsock pulled Karen out of the house. Then Mrs. Hartsock threw herself on Karen to put out the flames. Karen struggled to get up. She still thought Loretta needed her help.

The family held Karen down while fire swallowed up the house. As they stood there, the Hartsocks began to cry. Karen was badly burned. It seemed certain she would die.

Saving Karen

After several minutes, an **ambulance** arrived. It rushed Karen to the hospital. The doctors there were not hopeful. They said Karen had only a ten **percent** chance of living. Her burns were deep. They covered most of her body.

The doctors worked hard to save her. They put a tube in her throat to help her breathe. They removed the burned skin and replaced it with healthy skin.

They wrapped her in bandages. For many days, Karen hung between life and death.

But Karen **clung** to life. Her family begged her not to give up. And she didn't. Although the medicines made her sleepy, Karen struggled to stay awake. The **treatments** were painful, but she didn't complain about them.

Karen couldn't talk because of the tube in her throat. She couldn't see because of the bandages over her eyes. But she managed to signal, "I'm fine," and, "Thank you," to the surprised nurses.

One day Karen seemed very upset. She signaled and pointed. But no one could understand what she wanted. Finally, they figured out that Karen wanted to see Loretta. She was afraid Loretta had died in the fire and no one had told her.

In a few minutes, Loretta was sitting beside her.

"Please get well, Big Sister. We're all so lonely without you," Loretta said. Karen beamed. She started to improve soon after that.

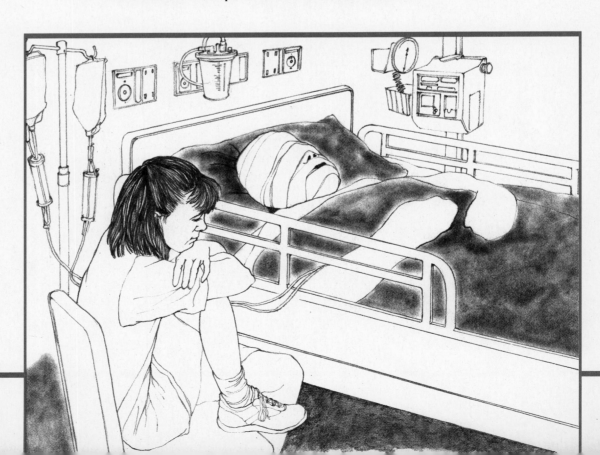

Finally, the danger passed. Karen would live. But her face and body were covered with thick **scars**.

When the bandages were removed from her eyes, Karen was shocked. She had not realized how different she looked. She began crying.

Karen was sad for a long time. But slowly she accepted her looks. She had many more operations. The doctors were able to replace some of the scar **tissue** with healthy skin.

Finally, after nine months, Karen was released from the hospital. She was very happy to go home.

In July, 1983, Karen won the Young American Medal for Bravery. She also won the Carnegie Medal for **heroism**. But Karen didn't see herself as a hero.

She said, "When you love the people in your family, you will do anything for them. That night I saw they were upstairs. I knew I loved them. I couldn't let them die. So I went back and back again for them. If you love somebody, you can do things you never dreamed you could do."

Do You Remember?

■ Read each sentence below. Write **T** if the sentence is true. Write **F** if the sentence is false.

_____ 1. Norma Hartsock started the fire.

_____ 2. Mr. Hartsock had just had an operation on his heart.

_____ 3. Karen's parents were out of town when the fire started.

_____ 4. Karen's sister Loretta was killed in the fire.

_____ 5. Karen saved her brother Johnny.

_____ 6. Karen's hair caught on fire.

_____ 7. Mr. Hartsock begged Karen to rescue Loretta.

_____ 8. Doctors were not hopeful about Karen's chances of living.

Critical Thinking — Main Ideas

■ Underline the two most important ideas from the story.

1. Norma was eleven years old.

2. An ambulance rushed Karen Hartsock to the hospital.

3. Karen wrapped Johnny in a blanket.

4. Karen Hartsock risked her life to save her brother and sisters.

5. Karen lived even though doctors did not think she would.

Exploring Words

■ Use the clues to complete the puzzle. Choose from the words in the box.

especially
cerebral palsy
treatment
ambulance
percent
clung
operation
scars
tissue
heroism

Across
1. a layer of cells
3. marks caused by injuries
5. something done using instruments to repair an injury
7. courage
9. a muscle disease
10. parts in each 100

Down
2. in a special way
4. takes people to the hospital
6. medical care given to a person
8. held on tightly

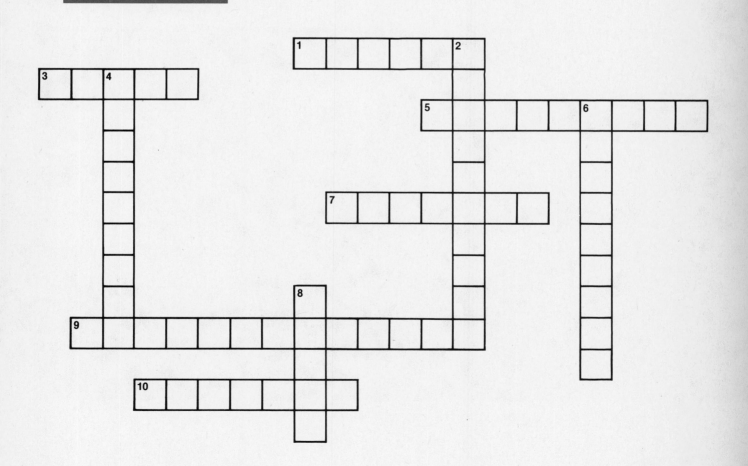

Five skydivers form a circle in the sky.

Free Fall Rescue

Debbie Williams was looking forward to Easter weekend. She and some friends would take part in a parachute **convention**. More than 420 jumpers would come from all over the country. They would meet in Coolidge, Arizona. There they would **skydive**. Williams couldn't wait. She loved the excitement of flying through the air like an eagle.

A Bad Feeling

Williams knelt in the huge airplane **hangar** in Coolidge. She was trying to pack her parachute. But the lines were tangled. She struggled to get them into her chute.

Gregory Robertson, a skydiving teacher, watched her. Something about the way she was packing the parachute bothered him. He walked over to her and introduced himself.

"How many jumps have you made?" he asked.

"Fifty-five," she answered.

Williams told Robertson that she was getting ready to do a six-way jump. For someone without much experience, this was a difficult jump. Robertson decided to keep an eye on her.

Six-Way Circle

The six-way jump called for six skydivers to work as a team. The first four jumped together. They held hands to form a circle. By spreading their arms, **arching** their backs, and bending their legs at the knees, they could keep their speed down to 120 miles per hour. The fifth jumper had to catch up with the circle. This diver would go into a dive that would allow him or her to fall quickly. The sixth

jumper had the hardest job of all. He or she had to dive very fast to catch up with the circle.

Later that afternoon, April 18, 1987, the plane took off. It carried 90 jumpers. Robertson told each person when to jump. Finally, everyone had jumped but Williams and her friends. Robertson would follow close behind this last group.

"Six-way!" Robertson shouted.

Out of Control

The jump began without problems. The plane leveled off at 13,500 feet. The six skydivers left the plane and began their **free falls**. Robertson also jumped out of the plane. He planned to dive slowly. He wanted to stay above the others and make sure everything went smoothly.

The first four people jumped holding hands. They would form the base. The fifth jumper left the plane. Robertson was surprised to see that Williams was the last jumper. She should have been one of the first four, he thought.

The first four jumpers had trouble forming the circle. Guy Fitzwater, one of the first four jumpers, ended up 500 feet above the circle. He could not catch up with the others. The fifth diver raced to join them. But he came into the circle too quickly. He knocked the others out of control. Debbie Williams circled above the group. She was about 20 feet above Guy Fitzwater.

Robertson watched what was happening. He knew Williams and Fitzwater were in trouble. If someone in the circle opened a chute, Williams or Fitzwater might fall into it. Then both jumpers might crash to the ground.

Robertson moved his body so that he was diving more quickly. He wanted to reach Williams and lead her away from the danger. But he didn't move fast enough. Williams went into a very fast dive. She was trying to catch the circle. As Robertson watched, she turned right into Fitzwater's path.

They're going to **collide**! Robertson thought with horror.

Miracle in the Sky

A moment later, 9,000 feet above the ground, Williams crashed into Fitzwater. She hit him hard with her face and chest!

Robertson quickly studied both jumpers. Fitzwater was hurt but able to control his fall. Williams, however, was knocked out. She rolled onto her back and fell straight down, like a rag doll. Her speed picked up to 165 miles per hour.

Robertson watched her for just a split second. He thought, I wonder if I can do this. Then he went into a sharp dive. With his head toward the ground, he pointed his toes and put his arms by his sides. He bent his head into his chest. Soon he was going 180 miles per hour. As he drew near Williams, he could see that her face was covered with blood.

When Robertson caught up with Williams, they were only 2,500 feet above the ground. In just fifteen seconds, they would hit the earth. Robertson knew he had only one chance. He had to get to Williams' **ripcord**. If he could pull that, her chute would open. That might slow her down enough to save her life.

Desperately, Robertson grabbed at the cord. When he caught it, he pulled hard. The chute opened.

Robertson then opened his own chute. He was

Gregory Robertson

just 2,000 feet from the ground. He had chased Williams for 25 seconds and 7,000 feet.

Robertson landed first. He was not hurt. But when Williams landed, she landed on her back. At first she didn't move. People on the ground rushed to help her. Her **skull** was cracked. She had eight broken ribs and a broken collarbone. A lung and kidney were bruised. But she was alive.

Gregory Robertson became a hero. President Reagan wrote him a letter. The press called him Superman. At the airport, his friends hung up a sign. It said In **Appreciation** of Skydiver Gregory Robertson Who Saved a Life on April 18, 1987. Good Job, Gregory! More than 400 people signed the banner.

Robertson didn't want people to make a fuss. He said, "I don't want to be a hero. I just want to be a skydiver."

But Robertson had performed an amazing rescue. Several other people who had tried such a rescue had not lived to tell about it. By risking his life, he saved Debbie Williams from certain death.

Do You Remember?

■ Read each sentence below. Write **T** if the sentence is true. Write **F** if the sentence is false.

_____ 1. Debbie Williams had never skydived before.

_____ 2. Gregory Robertson was the fifth jumper.

_____ 3. Williams crashed into Fitzwater in midair.

_____ 4. Williams and her group were the only people in the plane.

_____ 5. Robertson was a beginning skydiver.

_____ 6. Robertson pulled Williams' ripcord.

_____ 7. Williams landed without injury.

_____ 8. Robertson died when he hit the ground.

Express Yourself

■ Would you like to go skydiving? Tell why or why not.

_____

_____

_____

_____

_____

_____

_____

_____

Exploring Words

■ Choose the correct word from the box to complete each sentence.

hangar	desperately	skull	free fall	ripcord
appreciation	collide	skydive	convention	arching

1. The string which opens a parachute is the _____.

2. A building where airplanes are stored is known as

   a _____.

3. When two people or things knock together, they _____.

4. If you _____, it means you jump from a plane and

   perform a dive.

5. The other skydivers showed their _____ by making

   a banner for Robertson.

6. To fall through the air without an open parachute is

   to _____.

7. When you bend backwards, you are _____ your back.

8. The bone inside your head is called your _____.

9. If you feel hopeless and act in a careless way, it might be said

   you are acting _____.

10. Debbie Williams was looking forward to the parachute meeting

    or _____.

Trapped in a Well

Reba McClure stood in the bright Texas sunshine. She smiled at the group of children playing in the back yard. One of the children was her own 18-month-old Jessica. Reba loved to watch Jessica play with the other children.

Just then, at 9:30 A.M. on October 14, 1987, the phone rang. Reba ran inside to answer it. She was gone only a few minutes. But when she returned, the children were no longer playing. They were gathered around an eight-inch hole in the ground.

I Can't Let My Baby Die!

Reba looked for Jessica but did not see her. Her heart began to pound. She ran to the spot where the children stood. That was when the **nightmare** began. Reba learned that Jessica had fallen into an old, dry well.

The 17-year-old Reba became **terrified**. She ran inside and called the police. Then she ran back to the well. Three minutes later, the police arrived.

"She's here in back! She fell down right there!" Reba yelled to the officers. She stood over them as they checked out the hole. "I can't let my baby die! I've got to get her out."

Officer B. J. Hall tried to calm Reba down.

"We won't let her die," he said. But Hall was not sure he could keep that promise. Jessica was trapped about 22 feet underground in a space that was only 12 inches wide. Hall couldn't see that far down. He didn't know if Jessica was badly hurt. He didn't know if she had enough air to breathe. He wasn't really even sure she was alive.

The officers dropped a **microphone** down the hole. At first they heard nothing. Then they heard a sad little cry. It was Jessica. She was alive! Hall quickly ordered some equipment so they could start digging the child out.

Blasting Through Rock

City workers rushed in with a **backhoe**. They planned to dig a second hole right next to Jessica. They would dig 29 feet under the ground. Then they would dig a slanted tunnel up toward Jessica. This tunnel would break through the wall of the well about two feet below her. Then they could slide her down the well and into the new hole.

The workers dug down two feet. Then they hit rock. Bigger, heavier equipment was brought in. They struggled to blast through the **solid** rock. But **progress** was slow.

Hour after hour they worked. Their **drill bits** kept breaking on the rock. But they simply replaced them and kept on drilling. They stopped only to listen for Jessica's cries. As long as they heard her, they knew she was still alive.

When darkness fell, the workers had only gone a few feet. New workers came in to continue the job. Meanwhile, Reba and her husband, Chip, clung to each other in fear.

By the next morning, drillers finally got the second hole deep enough. Then they began to dig the tunnel toward Jessica.

This was the most difficult part of the job. The rock was very hard. It took an hour to drill through one inch of rock. A worker had to lie on his stomach while he drilled with a 45-pound

Reba McClure listens, hoping to hear Jessica.

jackhammer. Even the strongest person could only handle it for 30 or 40 minutes at a time. Then he would be completely exhausted.

"The jackhammer would bounce off the rock. It was like hitting a piece of steel," driller Paul Wilhite said. It was also frightening to be in such a small hole that far underground. Said Wilhite, "It was scary. It was like being in a **grave**."

One driller collapsed in the hole. He had to be carried up. Others had to leave when the dust and dirt choked them. But each time a worker came out of the hole, another one climbed in. The workers weren't giving up. They could still hear Jessica. When she began singing her favorite Winnie-the-Pooh song, their eyes filled with tears.

The Nightmare Ends

Meanwhile, the hours slipped away. Another night came and went. By the morning of October 16, the workers were getting close. But by this time, Jessica had been in the hole 48 hours. She was weak and hungry. Her right foot was jammed up against her face. She didn't cry much anymore; she hardly made a sound. Workers feared she was dying. One police officer called down to her to see if she was still **alert**.

"How does a kitten go?" he asked.

"Meow," came the faint reply.

The workers kept blasting through the rock. By noon, they finally reached Jessica. But the tunnel that led to her was only a few inches wide. Rescue worker Robert O'Donnell could touch her. But he couldn't get her out.

Again the drillers went to work. They made the tunnel wider. At last, at about 6 P.M., it was wide enough. O'Donnell went back down the hole. He crawled across the tunnel to Jessica.

"Come on, Juicy," he said, calling Jessica by her family nickname. "Just stay calm. I'm going to get you out."

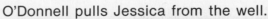

O'Donnell pulls Jessica from the well.

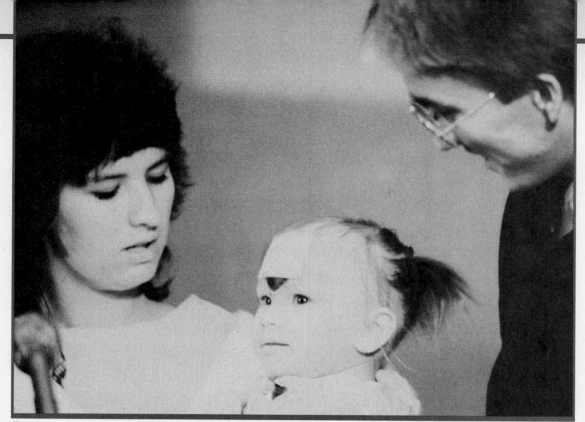
Reba, Jessica, and Chip McClure

Gently, O'Donnell pulled at Jessica. But she didn't move. She was stuck in the hole. He pulled her harder, and she began to cry. Still he kept pulling. For two long hours he worked to free her. He put clear jelly on the walls of the hole to help move her. Slowly, inch by inch, he pulled her closer. Then he gave a final tug, and Jessica fell into his arms.

"You're out, Juicy," he whispered. Doctors were afraid the fall might have caused broken bones. So O'Donnell strapped her to a board to protect her back. Then she was pulled up. Friends and reporters who had gathered around the hole cheered. Jessica had survived 58½ hours in the dark well. Her forehead was bruised, and her right foot was damaged. But for all she had been through, she was in good shape. Tears again filled the eyes of the workers who had struggled to free her. But this time they were tears of happiness.

Do You Remember?

■ In the blank, write the letter of the best ending for each sentence.

_____ 1. Jessica fell into an old
        a. house.   b. well.   c. car.

_____ 2. The backhoe only dug down two feet before hitting
        a. oil.   b. water.   c. rock.

_____ 3. A police officer asked Jessica to make the sound of a
        a. cow.   b. bear.   c. kitten.

_____ 4. Workers helped make a second
        a. hole.   b. drill.   c. jackhammer.

_____ 5. Jessica's nickname was
        a. Winnie-the-Pooh.   b. Juicy.   c. Reba.

Critical Thinking — Finding the Sequence

■ Number the sentences to show the order in which things happened in the story. The first one is done for you.

__1__ Reba ran inside the house to answer the phone.

_____ Jessica was strapped to a board to protect her back.

_____ Police officers arrived on the scene.

_____ O'Donnell put clear jelly on the walls of the hole.

_____ Drillers struggled to dig through the solid rock.

Exploring Words

■ Use the words in the box to complete the paragraphs. Reread the paragraphs to be sure they make sense.

nightmare	jackhammer	alert	progress	backhoe
microphone	grave	drill bits	solid	terrified

Reba McClure was **(1)** _____ to discover that little

Jessica had fallen into a deep hole. Police officers dropped a

(2) _____ into the hole so that they could talk to the

child. First, workers used a **(3)** _____ to try and dig her

out. But it could not dig through **(4)** _____ rock.

Workers also brought in a **(5)** _____, but still

(6) _____ was slow.

Hour after hour workers struggled to dig deeper. The

(7) _____ on their machines kept breaking. As the

tunnel got deeper, it became scary for the workers. Some of them

felt like they were working in a **(8)** _____. At last, they

reached Jessica. She was frightened but **(9)** _____.

When she was carried out of the hole, everyone cried with joy.

After 58½ hours, the **(10)** _____ was over.

The Little Boy Who Could

Kelley Lyons couldn't move. She lay trapped inside the wrecked pickup truck. Pain shot through her arms. The sharp edge of the door cut into her head.

Under Kelley lay her five-year-old son Rocky. He had been asleep when the accident took place. Slowly, he opened his eyes.

"Look, Mama," he said. "The car's upside-down, and the wheels are pointing toward the sky."

Kelley tried to look, but her face was covered with blood. Panic rose inside her. She thought she was blind.

Trapped in the Truck

It was early Halloween morning, 1987. Kelley and Rocky were returning home from visiting friends.

Kelley Lyons was not a **reckless** driver. She knew that this back road in western Alabama could be dangerous. In fact, her high school boyfriend had died in an accident on this road. That was nine years ago. But she still drove it slowly. She always took extra care going around the turns. This night was no different. But this night she didn't see the giant **pothole** in the road.

At 12:40 A.M., Kelley's truck hit the pothole. "It jerked the wheel right out of my hand," she remembered. A tire stuck in the pothole. That caused the truck to flip over. It bounced off the road and down a 20-foot slope. Kelley threw herself on top of Rocky to protect him. When the truck stopped rolling, it was upside-down. The roof of the pickup was bent into a V-shape. Kelley and Rocky were trapped inside.

As Kelley lay there, she thought about her husband Marty. He was an All-Pro defensive end for the New York Jets football team. He was in New York now, getting ready for a big game. He wasn't home to notice that she and Rocky hadn't arrived. She wondered how long it would be before someone realized that they were missing.

Suddenly, Kelley smelled gasoline. She feared the truck would burst into flames.

"You've got to get out of here," she told Rocky. "You've got to get out of the truck."

Rocky nodded. He began to **wiggle**. He managed to get out from under his mother. He crawled out the truck window. Then he came back and reached for Kelley.

He tried to pull her out. But it was no use. Rocky weighed only 55 pounds. He stood four feet two inches tall. Kelley weighed 50 pounds more than her son and was more than a foot taller.

Rocky decided to try something else. He crawled back into the truck and got behind Kelley. He pushed until he got her out of the truck.

At the Bottom of the Ravine

Kelley was **relieved** to be out of the truck. But she was still at the bottom of the **ravine**. She knew she was badly hurt. The blood in her mouth and nose was making it hard to breathe. She had to get help.

Rocky offered to climb up the bank and stop a car. But Kelley said no. It was too dark. It would be hard for drivers to see him. She was afraid he might be hit.

"Come on, Mom. I can help you get up the hill. I can push."

Slowly, they began crawling up the **steep** hill. Rocky got behind Kelley and pushed.

"Gosh, Mom, I bet you weigh a thousand pounds," he said.

You Can Do It

Inch by inch they moved up the steep ravine.

"I kept trying to dig my fingers into the ground and pull myself up by the roots of the grass," Kelley remembered. She felt terrible shooting pains in her arms and shoulders. "I didn't think I could do it. It hurt so bad," she said. By the time they got halfway up the hill, she was ready to quit.

"I can't go any further," she whispered. In the darkness, Rocky hugged his mother. He remembered a children's story Kelley often read to him. It was The Little Engine That Could. In the story, a little train tries to climb a huge mountain. It looks impossible. But the train keeps saying, "I think I can, I think I can."

"Mommy, you've got to remember the train. You know, 'I think I can, I think I can.'"

Kelley thought about what Rocky was saying. He was right. Climbing the hill was their only hope. She couldn't give up now. Slowly, she began struggling up the hill again.

"You can do it, you can do it," Rocky said.

His words echoed in her head. They forced her to keep going.

At last Rocky and Kelley reached the top of the ravine. As a car passed by, its lights lit up Kelley's face. Rocky saw how badly his mother was bleeding. For the first time since the accident, he began crying.

In a few minutes, a truck passed them and then turned around and came back.

"My mom's hurt real bad," Rocky told the driver. The woman drove them to the nearest hospital.

There the blood was washed from Kelley's face. She discovered that she was not blind. But her face was badly cut. The bones in both shoulders were **shattered**. Doctors bandaged her arms. They performed an eight-hour operation to repair her face. It took over 200 **stitches**.

The hospital called Marty Lyons, Kelley's husband, to tell him of the accident. He flew home immediately. Doctors told him that Kelley would be all right.

"If anything had happened to Kelley and Rocky, I would have hit bottom," said Marty.

Both Kelley and Marty are very proud of their son. Kelley is sure that she would have **bled** to death if Rocky had not been there to help her. His **courage** helped save her life.

Do You Remember?

■ In the blank, write the letter of the best ending for each sentence.

_____ 1. The accident occurred in
   a. New York.   b. Alabama.   c. Hawaii.

_____ 2. Kelley was afraid the truck was going to
   a. burst into flames.   b. sink.   c. get wet.

_____ 3. Rocky pushed his mother
   a. into a river.   b. up a hill.   c. to a church.

_____ 4. Rocky remembered a story about a little
   a. circus elephant.   b. mouse.   c. train.

_____ 5. Kelley feared the accident had damaged her
   a. eyes.   b. truck.   c. suitcase.

Express Yourself

■ Pretend you are Rocky Lyons. Tell a friend about the accident. How did you feel? What did you do?

_____

_____

_____

_____

_____

_____

_____

_____

Exploring Words

■ Read each sentence. Fill in the circle next to the best meaning for the word in dark print. If you need help, use the Glossary.

1. The truck hit a **pothole**.
   ○ a. hole in the road   ○ b. cooking pan   ○ c. tree

2. Rocky started to **wiggle** out of the truck.
   ○ a. tickle   ○ b. laugh   ○ c. twist

3. Kelley was **relieved** to be out of the truck.
   ○ a. sad and depressed   ○ b. happy   ○ c. worried

4. The truck rolled down into a **ravine**.
   ○ a. street   ○ b. open meadow   ○ c. ditch

5. Kelley and Rocky climbed the **steep** hill.
   ○ a. almost straight up and down   ○ b. rocky   ○ c. muddy

6. Rocky showed **courage** in helping his mother.
   ○ a. silly thoughts   ○ b. good manners   ○ c. bravery

7. The accident **shattered** several of Kelley's bones.
   ○ a. colored   ○ b. broke   ○ c. froze

8. Kelley's face **bled** from the many cuts.
   ○ a. blushed   ○ b. lost blood   ○ c. broke

9. Doctors put many **stitches** in Kelley's face.
   ○ a. sewing loops   ○ b. baths   ○ c. blankets

10. Kelley was not a **reckless** driver.
    ○ a. perfect   ○ b. damaged   ○ c. careless

A Path to Freedom

R oy Ahmaogak **zipped** across the snow on October 7, 1988. He was riding his snowmobile near Point Barrow, Alaska. Suddenly, he saw a whale 100 yards offshore. He stared. No, it was three whales! Their **snouts** were sticking out of a small hole in the ice. Ahmaogak was surprised to see them. All the other whales had gone south for the winter.

As Ahmaogak watched, the gray whales **heaved** their noses up through the hole in the ice. They **gasped** for air. Then they disappeared under the ice again.

No Way Out

When Ahmaogak got home, he told some friends what he had seen.

"Those whales won't last long," said one man.

"That's right," said another. "The ice has closed in on them. They're trapped. They'll be dead in just a day or two."

The **local** villagers knew the whales' breathing hole would grow smaller and smaller. Soon it would be covered with ice. Then the whales would have no way to breathe. The whales would have a chance to survive if they could make it to open water. Open water was five miles away. But those five miles were covered with ice that was two feet thick. The whales couldn't swim that far without coming up for air. They would drown.

The three whales came up for air every four minutes. They poked their noses up through the hole. But it was getting harder for them to do. The hole wasn't big enough. The **jagged** ice cut their skin. Their noses and heads became bloody.

The villagers decided to do something to try to help the stranded whales. People hammered away at the edges of the hole all night. They kept it from closing up with ice. But they knew that to save the trapped animals, more help was needed.

More Help Arrives

The story of the three whales spread across the country. **Volunteers** hurried to Point Barrow. They brought **chain saws**. They tried to cut a path to the open water. But the ice was too thick.

Rescue workers brought in a helicopter. It carried a five-ton steel block. When the block was dropped, it made a hole through the ice. But it took time to make the holes. And time was running out.

Two men from Minnesota flew to Point Barrow with special machines that kept ice from forming. Their machines kept the breathing hole open.

By the end of the first week, the workers had given the whales names. They were called Putu, Siku, and Kanik. These were Eskimo names meaning Ice Hole, Ice, and Snowflake.

Free at Last!

Day after day the rescue continued. The Coast Guard came to help. **Scientists** flew in. Even President Reagan offered to do anything he could.

Still, the whales remained trapped. Twenty-four holes were cut along a half mile stretch of ice. It was hoped that the whales would use the path to move toward the open water. They did, but rescue workers still had four-and-a-half miles to go.

Then the Soviet Union offered to send in two icebreakers. These giant ships could grind through ice. Workers quickly accepted the offer. The icebreakers began speeding toward Point Barrow.

After struggling for two weeks, Kanik, the smallest of the three whales, stopped coming up for air. Putu and Siku were still alive, however. Every four minutes they **wearily** pushed their noses to the sky.

At last the Soviet icebreakers arrived. On October 27, they cleared the four-and-a-half mile path to open water. Workers cheered wildly. With tears of joy, they waved good-by to their two tired friends. Putu and Siku took one last gulp of air. Then they swam out of the ice and toward the freedom of the ocean.

Soviet icebreaker arrives.

Do You Remember?

■ In the blank, write the letter of the best ending for each sentence.

_____ 1. On October 7, Roy Ahmaogak spotted three
   a. snowmobiles.   b. whales.   c. seals.

_____ 2. Every four minutes the whales needed
   a. food.   b. fresh air.   c. rainwater.

_____ 3. Workers gave the whales
   a. names.   b. medicine.   c. warm baths.

_____ 4. Two icebreakers were sent in from
   a. Minnesota.   b. the Soviet Union.   c. Japan.

_____ 5. On October 27, Putu and Siku
   a. were freed.   b. died.   c. stopped eating.

Critical Thinking — Fact or Opinion?

■ A **fact** can be proven. An **opinion** is a belief. Opinions cannot be proven.

Write **F** before each statement that is a fact. Write **O** before each statement that is an opinion.

_____ 1. Whales have to come out of the water to breathe.

_____ 2. Snowmobiles should not be allowed out on the ice.

_____ 3. The whales were five miles from open water.

_____ 4. People across the country heard about the whales.

_____ 5. TV stations need to be more careful about what stories

   they report.

_____ 6. People should never let whales die.

_____ 7. Workers tried to cut through the ice with chain saws.

_____ 8. Whales are the most amazing creatures in the world.

Exploring Words

■ Use the clues to complete the puzzle. Choose from the words in the box.

zipped
snouts
heaved
gasped
local
jagged
volunteers
chain saws
scientists
wearily

Across

1. people who study nature
4. tried hard to raise
5. people who work without being paid
9. in a tired way
10. having rough, sharp edges

Down

2. power saws
3. noses
6. from a certain area
7. breathed with great difficulty
8. moved quickly

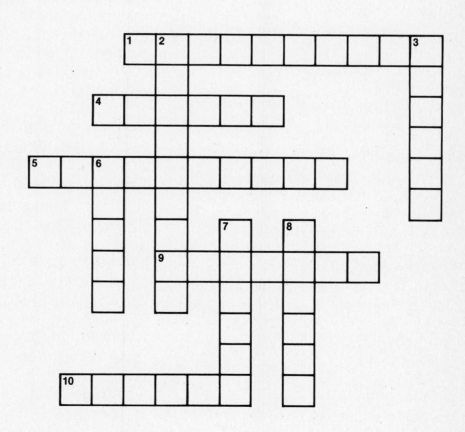

Glossary

abandon, page 38
To abandon means to leave and not return.

admiral, page 37
An admiral is a navy officer.

alert, page 68
Alert means wide-awake.

altitude, page 29
Altitude is the height above a certain level.

ambulance, page 51
An ambulance is a type of car that takes sick or hurt people to a hospital.

antitoxin, page 9
Antitoxin is a medicine used to cure diphtheria.

appreciation, page 61
Appreciation means being thankful for something.

arch, page 58
If you arch your back, you curve it.

assistance, page 3
Assistance is giving help.

backhoe, page 66
A backhoe is a tractor with a jaw-like basket that is used for digging.

bled, page 77
Bled means to have lost blood.

bobbing, page 18
Bobbing means moving up and down.

buried, page 25
If something is buried, it is covered.

calm, page 24
Calm means not upset.

cerebral palsy, page 50
Cerebral palsy is a disease that causes a lack of muscle control.

chain saw, page 82
A chain saw is a power saw. It has cutting teeth that are fastened to a moving chain.

cigarette, page 2
A cigarette is a roll of tobacco wrapped in paper for smoking.

clerk, page 43
A clerk is a person whose job is to keep records.

clung, page 52
Clung means to have held on tightly.

collapse, page 13
If you collapse, you fall down.

collide, page 60
If two things collide, they run into each other.

contact, page 32
If you try to make contact, you try to get in touch with someone.

convention, page 57
A convention is a meeting.

courage, page 77
Courage is the feeling that makes a person able to face danger.

cramp, page 32
If a muscle cramps, it tightens painfully.

cruise, page 17
A cruise is a ride taken on a boat for fun.

current, page 17
A current is the fastest moving part of a river.

death, page 16
Death means the end of life.

delay, page 25
Delay means to stop for a short time.

desperately, page 60
Desperately means being almost beyond hope.

diphtheria, page 8
Diphtheria is a disease that spreads easily from person to person. It makes breathing difficult and can cause death.

disaster, page 37
A disaster is something that happens suddenly and causes suffering and loss.

disease, page 9
A disease is a sickness.

dreadful, page 4
Dreadful means very bad.

drill bits, page 66
Drill bits are the cutting parts of a tool. They make holes.

equipment, page 3
Equipment is the supplies needed for a special job.

especially, page 50
Especially means in a special way.

exhausted, page 33
Exhausted means very tired.

experience, page 4
An experience is something that a person sees, does, or lives through.

explosion, page 2
An explosion is a sudden and noisy burst.

explosive, page 4
An explosive is the material used to blow things up.

fever, page 8
If you have a fever, your temperature is above normal.

flinch, page 5
To flinch is to back away from danger.

free fall, page 59
A free fall is when someone is falling through the air without his or her parachute being opened.

frostbite, page 30
If you have frostbite, part of your body is frozen.

fuel, page 37
Fuel is something that can be burned to make heat or power.

gasp, page 80
To gasp means to breathe with great difficulty.

grave, page 67

A grave is a hole in the ground for burying a dead body.

hangar, page 58

A hangar is a place to store and repair airplanes.

heave, page 80

Heave means to try hard to raise.

helicopter, page 37

A helicopter is an aircraft without wings. It is kept in the air by propellers.

heroism, page 53

Heroism means showing great courage when in danger.

horror, page 43

Horror is great fear or shock.

hostile, page 30

Hostile means unfriendly or uncaring.

hull, page 5

A hull is the body of a ship.

injure, page 44

To injure means to cause harm.

jackhammer, page 67

A jackhammer is a hand-held machine used to drill through rock.

jagged, page 82

Jagged means having sharp and rough edges.

jammed, page 43

If something is jammed, it is crowded.

life preserver, page 19

A life preserver is a ring that keeps a person floating in water.

local, page 81

Local means from a certain place.

microphone, page 66

A microphone is used to make sounds louder.

miner, page 22

A miner is a person who works in a tunnel under the ground. Miners dig minerals out of the earth.

miracle, page 25

A miracle is something very unusual and wonderful that cannot be explained.

mist, page 18

A mist is a very light rain.

musher, page 10

A musher is a person who drives a dogsled.

nightmare, page 65

A nightmare is a bad dream.

off-duty, page 44

Off-duty means to be off work. An off-duty pilot is a pilot who is not working at that time.

official, page 9

An official is the person in charge of something.

operation, page 49

An operation is something done using instruments to repair an injury.

panic, page 36

To panic means to have great fear.

percent, page 51

Percent means parts in each hundred. Ten percent means ten of each hundred.

plunge, page 43

Plunge means to dive quickly.

pothole, page 73

A pothole is a deep hole in a road.

powerful, page 19

Powerful means very strong.

progress, page 66

Progress means the amount of work done. If progress is slow, the work is going slowly.

rage, page 37

To rage means to continue out of control.

ram, page 5

To ram is to run into something with great force.

ravine, page 75

A ravine is a deep, narrow valley.

reckless, page 73

Reckless means careless.

relay, page 10

A relay is a race where each person on a team completes a part of the race. Usually an object is passed from person to person.

release, page 43

If you release someone, you set them free.

relieved, page 75

Relieved means to be free of worry; happy.

rescue

Rescue means to save from danger.

ripcord, page 60

A ripcord is the string that is pulled to open a parachute.

risk, page 11

Risk means to take the chance of doing something dangerous.

runway, page 42

A runway is a strip of ground used for the landing and takeoff of airplanes.

sandbar, page 16

A sandbar is a ridge of sand formed in the water by tides or currents.

scar, page 53

A scar is a mark left on the skin after an injury has healed.

scientist, page 83

A scientist is a person who studies nature and other areas of science.

sense, page 42

If you sense something, you feel it.

shatter, page 77

If something shatters, it breaks into pieces.

shock, page 19

Sometimes a body shuts down after an injury or accident. This is called a state of shock.

skull, page 61
The skull is the bone that protects the brain.

skydive, page 57
To skydive is to jump from an airplane with a parachute.

slump, page 30
If someone slumps in a chair, he or she is half sitting and half lying down.

smashed, page 30
To smash means to break.

smother, page 37
To smother means to keep the air from something. They smothered the fire with blankets.

snout, page 80
A snout is the part of an animal's head that has the jaws, mouth, and nose.

solid, page 66
Solid means hard and strong.

SOS, page 24
SOS is a call for help.

spirits, page 24
If people try to keep their spirits up, they try to feel good.

spurt, page 37
To spurt means to pour out suddenly.

stale, page 24
Something that is stale is not fresh.

steep, page 75
A steep hill is almost straight up and down.

stitch, page 77
A stitch is made using a needle and thread. It sews things together.

struggle, page 19
To struggle means to try very hard. If the swimmers were struggling, they were trying hard not to go under the water.

stun, page 44
To stun means to surprise.

survivor, page 25
A survivor is a person who lives through something dangerous.

swollen, page 8
If something is swollen, it is larger than its normal size.

system, page 37
A system is a group of parts that works together.

terrified, page 65
To be terrified means to be very frightened.

throb, page 31
Throb means to beat hard or fast.

tissue, page 53
Tissue is a layer of one kind of cells.

treatment, page 52
Treatment is the medical care given to a person.

violently, page 22

Violently means with great force.

volunteer, page 82

A volunteer is a person who works without being paid.

waterproof, page 30

If something is waterproof, it will not let water through.

wearily, page 83

Wearily means in a tired way.

wiggle, page 74

Wiggle means to twist and turn.

zip, page 80

To zip means to move quickly.

Chart Your Progress

Stories	Do You Remember?	Exploring Words	Critical Thinking	Express Yourself	Score
A Ship in Flames			/////		/20
Dogsleds to the Rescue				/////	/22
Trouble on the Mississippi			/////		/23
Miracle at Springhill				/////	/23
Survival in the Yukon			/////		/20
Abandon Ship!				/////	/22
Pulled From the Potomac			/////		/20
Do or Die!				/////	/20
Free Fall Rescue			/////		/23
Trapped in a Well				/////	/20
The Little Boy Who Could			/////		/20
A Path to Freedom				/////	/23

Finding Your Score
1. Count the number of correct answers you have for each activity.
2. Write these numbers in the boxes in the chart.
3. Ask your teacher to give you a score (maximum score 5) for **Express Yourself.**
4. Add up the numbers to get a final score.

Answer Key

A Ship in Flames
Pages 2-7

Do You Remember? 1-b, 2-c, 3-a, 4-b, 5-c

Express Yourself: Answers will vary.

Exploring Words: 1. equipment, 2. assistance, 3. cigarette, 4. explosion, 5. hull, 6. dreadful, 7. explosives, 8. flinch, 9. experience, 10. ram

Dogsleds to the Rescue
Pages 8-15

Do You Remember? 1-T, 2-F, 3-F, 4-F, 5-T, 6-T, 7-F, 8-T

Critical Thinking — Cause and Effect: Answers may vary. Here are some examples.
1. he did not have any antitoxin and he was afraid the disease would kill everyone in town.
2. it was the closest town that had antitoxin.
3. of the high winds and cold temperature.
4. pieces of ice were stuck in their feet.

Exploring Words: 1-a, 2-b, 3-a, 4-a, 5-c, 6-c, 7-c, 8-b, 9-b, 10-a

Trouble on the Mississippi
Pages 16-21

Do You Remember? 1-T, 2-F, 3-T, 4-T, 5-F, 6-F, 7-T, 8-F

Express Yourself: Answers will vary.

Exploring Words: Across: 2. shock, 5. powerful, 8. currents, 9. mist, 10. bobbing
Down: 1. life preservers, 3. cruise, 4. struggling, 6. death, 7. sandbar

Miracle at Springhill
Pages 22-27

Do You Remember? 1-a, 2-b, 3-a, 4-b, 5-c

Critical Thinking — Fact or Opinion? 1-F, 2-O, 3-F, 4-O, 5-F, 6-F, 7-O, 8-F

Exploring Words: 1. stale, 2. SOS, 3. buried, 4. miners, 5. spirits, 6. survivors, 7. violently, 8. calm, 9. miracle, 10. delay

Survival in the Yukon
Pages 28-35

Do You Remember? 1-c, 2-b, 3-b, 4-b, 5-a

Express Yourself: Answers will vary.

Exploring Words: 1. altitude, 2. smashed, 3. slumped, 4. hostile, 5. waterproof, 6. frostbite, 7. throbbed, 8. cramps, 9. exhausted, 10. contact

Abandon Ship!
Pages 36-41

Do You Remember? 1-T, 2-T, 3-F, 4-T, 5-F, 6-T, 7-F, 8-T

Critical Thinking — Drawing Conclusions: Answers may vary. Here are some examples.
1. the fuel system started to leak and oil spurted onto a hot pipe.
2. the fire was out of control and would soon burn the entire ship.
3. it heard the Prinsendam's SOS.
4. some of them were older and had health problems.

Exploring Words: 1-a, 2-b, 3-b, 4-a, 5-b, 6-c, 7-a, 8-a, 9-c, 10-b

Pulled From the Potomac
Pages 42-47

Do You Remember? 1-c, 2-b, 3-c, 4-b, 5-b

Express Yourself: Answers will vary.

Exploring Words: 1. runway, 2. jammed, 3. plunged, 4. stunned, 5. horror, 6. off-duty, 7. injured, 8. released, 9. clerk, 10. sensed

Do or Die!
Pages 48-55

Do You Remember? 1-F, 2-T, 3-F, 4-F, 5-T, 6-T, 7-F, 8-T

Critical Thinking — Main Ideas: 4, 5

Exploring Words: Across: 1. tissue, 3. scars, 5. operation, 7. heroism, 9. cerebral palsy, 10. percent
Down: 2. especially, 4. ambulance, 6. treatment, 8. clung

Free Fall Rescue
Pages 56-63

Do You Remember? 1-F, 2-F, 3-T, 4-F, 5-F, 6-T, 7-F, 8-F

Express Yourself: Answers will vary.

Exploring Words: 1. ripcord, 2. hangar, 3. collide, 4. skydive, 5. appreciation, 6. free fall, 7. arching, 8. skull, 9. desperately, 10. convention

Trapped in a Well
Pages 64-71

Do You Remember? 1-b, 2-c, 3-c, 4-a, 5-b

Critical Thinking — Finding the Sequence: 1, 5, 2, 4, 3

Exploring Words: 1. terrified, 2. microphone, 3. backhoe, 4. solid, 5. jackhammer, 6. progress, 7. drill bits, 8. grave, 9. alert, 10. nightmare

The Little Boy Who Could
Pages 72-79

Do You Remember? 1-b, 2-a, 3-b, 4-c, 5-a

Express Yourself: Answers will vary.

Exploring Words: 1-a, 2-c, 3-b, 4-c, 5-a, 6-c, 7-b, 8-b, 9-a, 10-c

A Path to Freedom
Pages 80-85

Do You Remember? 1-b, 2-b, 3-a, 4-b, 5-a

Critical Thinking — Fact or Opinion? 1-F, 2-O, 3-F, 4-F, 5-O, 6-O, 7-F, 8-O

Exploring Words: Across: 1. scientists, 4. heaved, 5. volunteers, 9. wearily, 10. jagged
Down: 2. chain saws, 3. snouts, 6. local, 7. gasped, 8. zipped